Matter and Energy

Matter and Energy

✦

A One Particle Theory of Matter and Energy

Carol U. Beatty

iUniverse, Inc.

New York Lincoln Shanghai

Matter and Energy
A One Particle Theory of Matter and Energy

iUniverse, Inc.

For information address:
iUniverse, Inc.
2021 Pine Lake Road, Suite 100
Lincoln, NE 68512
www.iuniverse.com

ISBN: 0-595-31900-9

Printed in the United States of America

Contents

AUTHOR'S NOTE

MATTER AND ENERGY is about a particle and how it inter-relates with all physical matter. It is a one-particle theory of matter, energy, and the Universe "the Ultimate Mass Particle". You will find it full of new interesting exciting theories, with proofs, on how the Universe functions. A lot of work has been done on the four forces, Gravity, Electrical, the Weak Nuclear force and the Strong Nuclear force. Good theories will be provided on the mechanics of what makes the forces function as they do. An interrelationship between the forces has been accomplished. There has been much work done on the similarity between Gravity and Inertia. This effort has produced surprising results. There have also been surprising results on the work with Quantum Mechanics and reality. As the book progresses, all the forces will be combined into one. It may also be of interest to find the atom has been redesigned "again". There will be serious discussions and theories on how the mechanics of matter functions. How the **five states** of matter functions. There has been much work done on the matter-energy relationship. This book will provide interesting theories on how the Universe began and ends. In classical physics, the primary concentrations of energy in the Universe are found in large bodies of matter such as the sun. In this book most of the energy, as well as matter, is found elsewhere.

You will also find nature's wondrous secrets will no longer be secrets, for the very foundation of the "everything theory" lies within this book, *MATTER AND ENERGY.* If the theories in *MATTER AND ENERGY* are accepted they will have an earth-shaking effect on physical science and the entire scientific community. It's for real. This book builds upon its self; meaning there are many new theories and hypothesis that are interrelated. At the end of *MATTER AND ENERGY*, there is a listing of 9 rules used in developing this book. There are also answers to the **11 greatest unanswered questions of physics.** A pleasant easy comprehension will be accomplished best by reading it in its entirety, from beginning to end. I hope you find this readable and understandable book as interesting and exciting as I do.

MATTER AND ENERGY

This book "Matter and Energy" is a one-particle theory of Matter Energy and the Universe. It is a well-thought-out theory based on the hypothesis that there is only one mass particle that all matter is composed of. This particle is called the *Ultimate Mass Particle* or UMP. In an attempt to prove the existence of these particles, Matter and energy will discuss how the composition of all forms of matter is accomplished with only the existence of one kind of particle. We will probe deep into nature and its forces. A thorough discussion of the *whys* and *hows* of all the forces will be provided. We will work with how all the forces and phenomena in nature are propagated by the particle and hopefully you will finish Matter and Energy believing that the particle is capable of doing what this book says it dose. If nothing else, you will surely find this way of thinking of our physical Universe to be indisputably new. The particle itself is not new, for it has been around since the beginning of time. The Ultimate Mass Particle is in fact an **electron,** a **free space electron.** The writer of this book does not believe that this tiny little mass point, the electron, has been understood well. Although it is known to be involved in most physical phenomena, what it is and does has not been understood completely. In Matter and Energy we will try to get better acquainted with the **electron**, or **UMP**, whichever you wish to call it. To keep things straight, we will repeat the fact that the UMPs are the Universe.

Matter and Energy is about our Universe in which all physical phenomena can be explained. You should find this theory of matter and energy vastly different from the way you have been thinking of nature's forces acting and reacting with each other. It will be observed that all the forces react with each other without any special set of arbitrary rules to make them function as they do. All the forces have been united into one. You will find they work together in complete harmony.

It's time to get started and see what this Universe is really made of. We will find that it consists mostly of free space, free space meaning all the space in the Universe that is not occupied by some form of matter. Free space also, therefore, refers to all space and the voids within the atoms and molecules. This space is filled with inertial energy carrying particles, better named the free space electron. These particles are in constant motion with a velocity greater than the light speed motion of the light waves they carry. They are also in a uniform, random, disorder and are in constant collision with each other throughout the Universe. These particles account for all forms of mass, energy, and all the forces.

This Theory is a one-particle theory of matter energy and the Universe. These particles and the inertial energy they possess in their velocity and spins accounts for most of the energy in the entire Universe. The only energy matter possess is the kinetic energy of their own motion. The entire Universe is mechanical or kinetic. Everything that happens in the Universe is a direct or indirect result of this particle force. Inertial energy is the only form of energy in this Universe. Matter in itself consists entirely of these particles in a near rest state. These free space high-energy electrons hold and force atoms, molecules, neutrons, protons and all matter together. The free space electron possesses no energy and cannot receive or give up energy except that of its own inertial energy, which consists of linear and spinning motion. Its spin and velocity governs all physical phenomena and generate all the forces. This concept of the Universe may never have been thought of before.

Now this is the Universe, theorized as it may be, if you can let your mind accept the following theories, we can take an interesting, hopefully exciting trip through the Universe, examining, predicting and understanding the mechanics of all the forces and phenomena within our Universe. This new concept of the Universe will have to live within the old laws of physics, so throughout this book there will be a constant comparison between new and old ideas.

Gravity

Gravity one of the four forces, should be a good place to build a better understanding of our Universe. We are more or less building a Universe, so gravity seems to be an appropriate force to work on first. This is because it affects all matter and has a great deal to do with how the Universe functions. We need to think of gravity a little differently then we have been. Gravity is generally thought to be an attracting force. The illusive graviton is thought to possess an attracting force between all forms of matter. We are so certain that gravity is an attracting force that we have written it into a law, the law of Universal Gravitation. It is very strange that a law of physics would be written with the arbitrary term "attract" written into it. There really is no more reason to think of gravity as being an attracting of pulling force than it is to think of it as being a repelling or pushing force. The fact is if we try to envision an attracting force between two particles of matter, we will experience great difficulty. You cannot really think of any way two objects can attract each other without a physical link between them. A chain, a rope, or bolts can pull objects together.

The truth of the matter is that there are no attracting forces in nature, **nor can the mechanics of such forces be explained**. On the other hand, repelling, or pushing forces, are easy to envision; simple things like throwing rocks at two objects will force them together. Blowing air pressure at the objects could make a repelling force that would move them together. If the objects move easily, sound waves could force them together. There are endless ways that matter can be forced or repelled together and really no ways that can be **thought** for them to be attracted together.

Even Under the right circumstances, electromagnetic waves can create the repelling force that moves particles of matter together. Electromagnetic waves are close to what gravity really is. The so-called Electromagnetic waves and gravity both react out into space at the speed of the light wave and their effectiveness is inversely proportional to the square of the distance. In Matter and Energy, gravity is **not** an attracting force. I may have to talk fast to put this over.

Now In our Universe there are no attracting forces. This is a very bold statement if for no other reason than every human on the planet disagrees with it. Gravity does not attract. Unlike poles of a permanent magnet do not attract. Unlike electrical charges do not attract. A vacuum does not attract. Oh, I have found one that we agree on. Check your memory. It was not all that long ago that a vacuum was thought to be an attracting force, and it is not too illogical for a vacuum to be thought of as attracting. Vacuum is a pressure difference. The vacuum side is a low-pressure area. The atmospheric air pressure side is the high-pressure area that makes objects move into the low-pressure area. The early thinkers or philosophers had a difficult time understanding the vacuum, air pressure relationship. The reason for their difficulty was that atmospheric air pressure seems to be everywhere around all pieces of matter making it difficult to detect.

Back to gravity and our particle. Gravity is not too different than the vacuum. In the vacuum, as you probably know, the rapid inertial motion of the molecules in the air, supply the motivating inertial force, or pressure, that will cause pieces of matter to move. That is, they move because there is a lesser molecular bombardment of air molecules on the vacuum side of the piece of matter than there is on the atmospheric pressure side. This is true for there are fewer air molecules to exert an inertial pressure on the vacuum side then there is on the air pressure side. The molecules are farther apart. A vacuum can also be created by allowing the air molecules to slow down losing some of their inertial energy "lose heat".

Now gravity has its free space electrons to do the work. The electrons also are everywhere. In fact, they are more everywhere than the air molecules. They fill all of free space, all the space inside of all particles of matter, inside the molecules, and even inside the atom. This makes them undetectable, almost. The only place the free moving, in constant motion, bombarding particles are not is inside the neutron and the proton.

These particles are the strongest force in the Universe, and gravity is one of the forces they propagate. These little mass points "free space electrons" are capable of exerting the tremendous pressure that will move any mass in the Universe. They do this with their inertia, the kinetic energy they possess in their light-speed motion. The little fellows are what make it all happen.

In gravity, this is done by lowering the particle pressure of the free space electrons in an area, and then the normal bombarding pressure of the free space particles will make all matter move in that direction. The pressure lowering happens as the free space electrons pass through ordinary matter such as our earth. As they pass through a portion of them collides with the nuclei in the various atoms they pass through. This causes the particles to slow down a little as they lose most of their linear motion as it is converted into various spins. The overall effect is a lowering of the particle pressure or their general slowing down. By slowing down, it is not meant that the particles lose energy. This means that the linear motion slows as the spinning motion speeds up. This decrease in some of the linear motion of the particles makes a less violent inertial reaction between particles and pieces of ordinary of matter. This less violent reaction when the particles collide is what we are calling a lessening of normal particle pressure. This allows all particles to be forced toward each other by the normal particle pressure in the free space electrons that are beyond the particles of matter. Gravity it is called.

In the above paragraph the description of inertial gravity might be easier to understand if it was thought of as a shadow effect. This shadow effect is where free space electrons, traveling towards the earth, have much linear motion "pressure" where as those leaving have less linear motion "less pressure". The free space electrons may travel at light speeds between collisions, however, they do not really go anywhere. It would be a slow, difficult process for free space electrons to actually enter the earth and come out the other side. It would be better to think of the earth as a barrier or shield. This means that the particle pressure of the free space electron leaving the earth surface and on out into space would exert much less pressure then the normal inertial effect of the free space electron. This would force all matter towards the earth "inertial gravity". This shadow or shielding effect is not only for large pieces of matter. The shadow effect also works on small things. Later in this book when we work on an atomic scale this shadow effect will be understood better.

Gravity is the force that pushes the moon towards the earth and keeps it from flying off into space. It also is the one that makes the red apple fall from a tree. This theory predicts that things happen more slowly near large pieces of matter such as the earth. This has been experimentally proven to be true. The reasoning behind this prediction is the free space electrons are moving a little slower near the earth and they are what propagate all physical phenomena, so consequently everything moves slower near large pieces of matter. The speed of light is also part of the "everything" that moves slower. This repelling force called gravity makes another prediction. This is how inertial gravity functions on a larger scale, such as the Universe. The prediction is that distant galaxies on the fringes of the Universe will be pushed away by the repelling force of inertial gravity. This is true, for the tremendous particle pressure of the free-space-electrons within the Universe are pushing very hard on these galaxies. This causes them to accelerate on into never-never land. Cosmologists have observed this repelling force of inertial gravity pushing the galaxies away. This is very **strong** evidence that inertial gravity is a repelling **force** that fills the **entire** Universe.

Another example of free space electron pressure is when two metal plates are placed very close together where they will experience an inertial force pushing them together. The shielding of shadow effect of the two metal plates provide an area between the plates where the free space electrons exert less inertial pressure then the electrons outside the plats. This forces the plates together. Analogous to the way molecules are held together by the inertial pressure of free space electrons. Quantum mechanics describes this phenomenon with the metal plats in a different way, called, the Casimere effect. Matter and Energy calls this phenomenon a simple natural action of the inertial energy carrying free space electrons. This is another **proof** that the free space electron exists.

It was stated above that free space electrons lose no energy when passing through large pieces of matter. This is not entirely true. In large pieces of matter, such as the planet earth, some of the kinetic energy, velocity of the particles, are given up to the molecules of matter within the planet. This means that there is a large increase in the molecular motion "heat" of the matter beneath the crust of the earth and a decrease in the velocity of the particles. The energy that the particles lose will eventually be radiated back out into space in the form of waves "heat" carried by the free space electrons.

Something very similar to this happens in our sun. The sun radiates waves in the free space electrons the same as our earth. Of course they are on a much grander scale. These waves, carried through space by the free space electrons, are what make life possible on earth. The waves from the sun are carried in a similar manner as those leaving earth. On the sun, the free space electrons, carrying tremendous quantities of kinetic energy, enter the sun from all directions and leave from all directions. They give up some of their energy to the molecules of matter within the sun, the same as they do on earth. The energy loss is even greater in larger bodies than it is on earth. This in turn, increases the inertial gravitational effect which forces the molecules closer together. Here there is much high-speed mechanical action between molecules and free space electrons. In this harsh environment, the protons and neutrons have difficulty maintaining their identity. The protons and neutrons that are affected most are those contained in the hydrogen atoms. Many of the hydrogen atoms are broken apart and form new atoms of helium. Some of the nuclear parts, such as neutrons, do not survive the transition of becoming helium atoms. They are smashed and broken up into their basic building blocks. These basic building blocks are free space electrons. These electrons greatly over populate the **core** of the sun. The entire sun as well as all space in the Universe is full of free space electrons. These electrons are moving beyond the speed of light and are capable of giving up tremendous quantities of inertial energy. The free space electron will expend their energy getting all of these newly liberated particles "in the sun" into motion, light-speed-motion, like the rest of our free space electrons.

The above paragraph was of course describing a thermonuclear reaction. You should have noticed the hydrogen atom had no energy within it's self. **All** of the energy in the reaction came from the free space electrons. This is true of all thermonuclear reactions. There is no way ordinary matter can store inertial energy in the amount of $E=MC^2$. Matter is as dumb as a stone and has no magical powers enabling it to store kinetic energy in this amount. All of the energy of fusion came from the **free space electrons.** The energy the sun takes from free space will be given back to free space in the form of outward moving particles and very energetic waves of free space electrons. These high-energy waves are of various frequencies, so consequently they are of various energy levels. All of this energy is carried away from the sun by the high-energy waves in the free space electrons. This is what keeps the earth from being a very large ice cube. Some of the larger particles of ordinary matter that are expelled from the sun leave at a high speed. It might be possible that these particles of matter could be accelerated to a very high speed as they ride the wave action in the free space electrons. This could account for how the **cosmic rays** travel through free space. It is unlikely the sun would have enough energy to accelerate these particles of matter to near the speed of the light wave, however there are energy sources in the Universe that would. One example of where this very energetic wreckage of ordinary matter "cosmic rays" comes from is a massive supernova explosion. This explosion creates very strong free space electron waves of all frequencies and would give the cosmic rays a free ride.

This energy that is carried from the sun to the earth needs to be examined a little closer. The physics books simply state that the sun radiates energy in discreet bundles called photons, which will carry energy to the earth at the speed of the light wave. The writer of this book does not believe that any thing, or any kind of matter, can move through free space from the sun to earth at the speed of the light waves. This means nothing, not free space electrons, not photons of light, not space ships or rowboats. "The one exception might be particles of matter riding the waves in the free space electrons". The photons of light do not exist! The only way energy can travel from the sun to earth is to create a wave disturbance in the free space electrons. These waves will travel to earth where the **inertial** energy carrying free space electrons will physically react with all forms of matter. By physically react it is meant that inertial energy carried by the free space electrons will bombard all solid matter they come in contact with, causing them to vibrate "heat". This is how energy is carried to earth, not by some magical vehicle called a photon. The free space electron travels through free space at light speed velocities for only very short distances before they collide with other particles. The average distance these particles travel before they collide with another particle is about the diameter of the Bahr atom. They really do not move around in space great distances. They do, however, continuously exchange inertial energy with each other in their light speed collisions. This is the same way electron currant is carried in a copper wire. The copper wire is full of free space electrons hammering each other with light speed collisions. This means that the wire is full of inertial energy the same as all of free space. The energy is there even if there is no electron current flowing in the wire. Applying inertial pressure to the electrons in one end of the wire will start a **wave action** of electrons to pass through the wire. This wave action is capable of doing work. This is how electrical current is carried in a conductor. The free space electrons in the wire move through the wire at only about one-inch a minute. However, they carry a wave disturbance at the speed of light. When this wave or waves of free space electrons reaches its destination, it is capable of doing work. The same is true in fiber optics. Here, the glass fiber is always full of free space electrons that are hammering each other with light speed blows. This is the same as they do in the copper wire. All we have to do is start a light wave in the free space electrons that are within the glass fiber. This wave will travel at the speed of light through the fiber. The wave of free space electrons in the glass fiber is also capable of doing work, the same as an electron current in a conductor. **The only difference in visible light and electron current is the frequency of the vibration**

in the electrons. Visible light comes in many frequencies "colors" enabling it to do a large variety of important jobs that are not possible with an ordinary electron current in a conductor.

In the above paragraph it was stated that the photon does not exist. It would seem appropriate to give a little more support to this statement. The photon was originally created by Einstein to explain the photoelectric effect. Before Einstein created the photon, it could not be explain how a light wave would instantaneously knock bundles of electrons from a photoelectric metal plate; so the photon was invented to do the job. There is a problem with photons continuously knocking bundles of electrons from a metal plate. The problem is where does the metal plate find all the electrons to create a continuous flow of electron current? Waves of photons continuously knocking electrons from a metal plate is not possible. The reason for this is, the photoelectric plate would quickly run out of electrons. Conversely, waves of free space electrons, continuously striking, entering and overpopulating the plate with electrons would buildup pressure causing a continuous flow of electron current! This is **strong evidence the free space electron exists and the photon does not!**

There is one more hurdle that the free space electrons must clear. This hurdle involves the motion of the earth around the sun. It is believed by most of the scientific community that there cannot be a dense population of particles in free space. Their thinking is that if there were the earth would slow down as it collided with them and eventually crash into the sun. Right here on this paper an attempt will be made to clear this hurdle. It would be helpful if you remember the free space electrons can move freely through molecules of matter of which the earth is made of. This is the way it is done. To start with the free space electrons have a means velocity, between collisions, that could orbit the earth 11 times a second. So the earth is being hammered hard from all directions. The leading surface of the earth crashes into the free space electrons as it orbits the sun. A large portion of the free space electrons will have to flatten out and move parallel to the leading surface of earth. This causes a venturi effect that lowers the particle pressure on the leading surface of earth. The normal particle pressure on the trailing surface will then push the earth around the sun. The particles coming from the sun have little linear motion and much mixed up spins. This lowers the particle pressure on the sunny side of earth, (gravity). This concentration of mixed-up spins drags on the earth making it spin as it orbits the sun. The normal particle pressure on the earth's surface facing away from the sun will push the earth toward the sun, neutralizing the centrifugal force effect. This is how the earth passes through the free space electrons that fill all of free space. It is known that the sun hurls huge quantities of electrons out into free space every second. These electrons and other particles known to be in the vicinity of the sun should slow the earth down and it would eventually crash into the sun if it were not for the venturi-effect. This is another **proof** that the electrons can exist in free space. It would be a real surprise to most, if it were found necessary for the free space electron to exist in-order for our planet to spin on its axes and orbit the sun in the manner it does now. Wouldn't you agree?

We have wandered off a little from the subject of gravity. However, all of this is related to gravity. Thinking of gravity in this manner may be hard for some to accept. I'll bet Isaac wouldn't be to happy if he knew that somebody was saying that he has been looking at the Universe up side down. By up side down it is meant that Newton thought of gravity as an attracting force instead of a repelling force. It is conceivable that great philosophers or men of science, such as Galileo and Newton, may have pointed our modern physics in the wrong direction. Possible if the early thinking people of science had dome just a little different thinking on how various scientific phenomena functions, "such as field theories, the unrealities in quantum mechanics, gravity and others" we might find ourselves believing the Universe functions in a **totally different manner.**

Well, how did I do? Did I talk you out of believing in gravity? It is a fair start in building a new way to think of our physical Universe, isn't it?

Now it's time to move on. We still have most of the Universe that needs to be better understood. If we are building a better understanding of the Universe, we will need matter. A better understanding of it will be tried now. It was stated earlier that the entire Universe is made up of free space electrons and nothing else. This particle has been shown to be an energy particle. Now an attempt will be made to show that **it is this particle that makes the free space electron, the nucleus of the atom, "protons and neutrons"** which is the major portion of matter in all pieces of matter.

Neutrons and protons are composed of tightly packed free space electrons, in a static state. These little balls of matter are held together by the bombarding pressure of the free space electrons. They do such a good job of holding them together that they are kept fairly consistent in size. If they try to build to a larger size, their geometrical configuration will be such that the free space electrons will chip away at them until they are the size of protons or neutrons. The same is true if they are too small. The neutron or proton will be unstable. They also cannot be held together easily by the bombarding pressure of the free moving particles that are outside of these balls of matter. Instead of holding them together, the free space electrons will chip away at them and eventually they decay back into individual free space electrons. There is a small difference physically in the neutron and proton. The diameters are the same, but the proton has fewer particles in it. Since the diameters are the same, and the neutron has more particles in it, obviously, the proton must have some voids in it, while the neutron probably does not. There is another difference between the neutron and proton. It is generally thought that the proton carries a positive charge and will attract a negatively charged electron. In this book, the proton seems to get along well with the free space electrons. This means that there are likely to be electrons moving slowly and spinning in the vicinity of the proton. This can only happen if the proton is porous. By porous it might be meant that the free apace electrons could work or ricochet their way through the proton and come out the other side Moving slowly and spinning. A more likely possibility is that the arrangement of the particles on the outer surface of the proton could cause the free space electrons to lose most of their linear motion as it is converted into spinning motion when it slams into the proton. This could cause a slow moving electron to stay around the proton awhile.

If there is a cluster of protons and neutrons, as in the nucleus of an atom, there will be a number of slow moving electrons in the vicinity of the cluster. The number of slow moving particles depends on the number of protons only. The neutrons do not encourage particles to stay close to them. The reason for this is the particles on the outer surface of the neutron are packed a little closer together "smother". The neutron is not rough like the proton. All the particles "free spacer electrons" that come slamming into the neutron ricochet away at essentially the same velocity they came in at. A very small portion of the velocity is given up to neutrons because the neutron has a mass that is over 1000 times greater than the free space electron.

The only thing these free space particles accomplish is to hold the neutron together. They do a good job of holding them together, for they are hammering away on them from all sides, all the time. The one exception is when two or more neutrons are touching each other. At the point where they are touching, there is no particle pressure; consequently, the neutrons are held "forced" together extremely well. It takes a very strong mechanical disturbance to separate them. This is another example of matter exerting its shielding or shadow effect. The protons and neutrons in the nucleus of the atom are forced and held together in the same manner.

Well, this is your atom. A tight cluster of tiny balls of neutrons and protons which are held together by free space electron pressure. The protons and neutrons, particularly the protons, repel each other with a very strong force until they are very close together. This, of course, is true due to the free space electron pressure. These tiny particles densely populate all space where they are constantly vibrating and colliding among themselves and any piece of matter in the vicinity. When there is any room at all between the protons and neutrons, the ping-pong action of the free space electrons will try to force them apart with a very strong force. In fact, there is a strong repelling force between protons until they are within one ten trillionth of a centimeter apart. At this distance, there is no longer room for the free space electrons to squeeze between the protons, so there is no particle pressure at this point. This means that the tremendous particle pressure of the free space electrons will smash and hold them together forever, that is, hold them together forever unless some mechanical force exerted on the nucleus is strong enough to knock them apart.

Within this cluster, the neutron is a tiny ball composed of about 1839 identical electrons. The neutron is also held together by the pressure of light speed free space electrons. The protons and neutrons are forced and held together in the nucleus of the atom by the same free space electrons. The number of slow moving electrons around the so-called nucleus is dependent on the number of protons in the cluster. That is what the atom looks like when you think of the Universe consisting of only matter and energy. I hope you like it.

When thinking of small things like atoms, or large things like the Universe, it should always be kept in mind that everything in the Universe is composed of only one type of particle; the electron.

A little more thought on matter and free space would be appropriate now. What is thought of as free space is full of free space electrons. If these particles do what this says they do, these little mass points must be in very rapid motion, colliding with everything there is to collide with, and very close together. This means that an area, say one cubic foot of free space may contain nearly as much matter as a cubic foot of iron has in all the protons and neutrons within it. This may seem more plausible if you remember that solid matter is composed of almost all-empty space with only a small amount of it filled with protons and neutrons. The block of iron has more total mass than the same volume of free space because there are no voids in the so-called nucleus of the atom and there are voids between the free space electrons. These same free space electrons outside of the iron mass also hold it together. If we accelerate this iron mass, it will send out a disturbance, or wave action, in the free space electrons. Also, at a high velocity, there is an increase in pressure on the forward side of the object that will cause the mass object to be distorted. The object will become shorter. Later in this book, the distortion will be better understood when we describe how molecules of matter are held at a point of equilibrium in solid matter.

As the speed of the mass object becomes greater the free space electrons begin having trouble ricocheting their way through the object. They begin to build up on the leading surface of the object. More speed and the whole object become heavily populated with particles. This increases the mass of the object! A little more velocity and the greater pressure on the forward side than there is on the reverse side of the nuclei in our mass object may cause them to fall apart and decay back into individual free space electrons. So don't try to travel too close to the speed of the light wave or you may fall apart. The only thing that can travel faster than the speed of the light wave is the *free space electron*. Individual particles colliding or vibrating to carry a wave action must travel at a greater velocity than the speed of the wave.

This might be a good time to get to know some of the rules or assumptions that are used in this article. First, the word **heat** is not used. Heat is described in another way, usually as an increase in molecular motion or an increase in a certain frequency of electromagnetic waves. **Friction** is another word not used. There is no friction in this Universe. It is generally thought that friction causes heat. In this Universe, there is on heat or friction, the only form of energy is mechanical, so naturally, when one particle of matter rubs against another, there is an exchange of kinetic energy of one form or another. The only way that this particle force is felt in this Universe is to reduce or increase the free space electron pressure in some area of space. One method of reducing the pressure was discussed in the section of this book under gravity. Others will be discussed later.

We will move on again. This time the subject will be what physics books call electromagnetic waves, some-times called the particle wave theory. In this theory, it is thought to be a wave action, carried through space that can be totally empty, with nothing there to carry the wave. Then, through some form of magic, it mysteriously changes back into a particle. A very strange electromagnetic wave theory! It will be done differently in Matter and Energy.

In this book the energy is carried by a wave action in the free space electrons. In fact, these particles are the waves. Earlier in this book it was stated that the free space electrons are in uniform, random disorder. Any time the uniformity of the particles is disturbed, they are capable of exerting a detectable force. We will try to demonstrate this with the use of visible light carried by the free space electrons. This particular light will be generated by a flow of free space electrons in the filament of a light bulb. The particle flow causes the molecules in the filament to vibrate "heat", which in turn causes the free space electrons in the space surrounding the vibrating filament to vibrate. This vibration or waves, which is induced in the free space electrons, will travel at light speed from its source. The free space electrons that originally received the vibration may not travel through space very far, but the wave effect, set up in the free space electrons, will travel far, and at 186,000 miles a second.

Infrared light is another wave that is generated from a vibrating molecule or atom. It is generally thought that the nucleus of an atom has a definite number of electrons orbiting the so-called nucleus. In this book, there are electrons around the nucleus, but they are not so firmly attached as is generally thought. The free space electrons are what hold the nucleus of the atom together, so there is much action in the vicinity of the so-called nucleus. There are clouds of slow-moving and spinning electrons around the nucleus, but they are not necessarily the same ones all the time. The electrons may start the wave disturbance, such as a spark jumping a gap in an electrical circuit, but the free space electrons were started in motion by a more massive piece of matter; an excited nucleus of an atom. Ok, the vibrating nucleus starts the free space electrons to vibrating, which causes a wave effect in the free space particles. This wave effect is what we call infrared light.

The primary difference between visible light and infrared is the speed of the vibration or the wave "frequency". Their frequency is a little slower than visible light. In all the free space electron waves the higher the frequency, the more energy they carry. When these infrared waves strike another mass object, they do it with enough force to start the molecules vibrating at the same frequency as the original ones were. It is generally thought that infrared waves transmit heat. In this Universe infrared waves are just another way mechanical energy is mechanically transmitted from one mass object to another.

In general, all free space electron waves are transmitted in the same way. Their primary difference is the frequency of the waves in the free space electrons. There is of course, a difference in the way they are generated, and a difference in what happens when received by pieces of matter. One difference in the way they are received is the radio wave. As for this wave, when it strikes a properly tuned antenna, the free space electrons enter the antenna, setting up a wave action of electrons in the antenna.

We could go on through the entire electromagnetic wave spectrum, but it's not necessary, for they all are the same, with only a difference in the frequency of the disturbance in the free space electrons. This disturbance is a variation in the normal particle pressure, which we call a wave effect.

It's very similar to the way sound is carried by air molecules. In air, the molecules in the immediate area of the original disturbance are given a push, causing them to become denser, or closer together which is more pressure. Its a series of high and low pressure that travels through air, making up sound waves in the atmosphere.

The same is true with the free space electron waves. If a filament, or other source, is vibrating, there will be a series of high and low pressure "densities" in the free space electrons, the same as sound in air molecules. This effect is capable of doing work. The means of detection and the type of work done may vary, but it is still a simple mechanical action being transferred from one object to another by a mechanical means.

I may have made the waves in the free space electrons sound too much like sound waves. They are different. For starters, the free space electron wave effect, which is carried by the free space electrons only, travels at the speed of the light wave. The speed of light waves varies a little near mass objects. It slows down, for example, near the earth. The speed also varies when passing through pieces of matter. Although most of the wave carries by the free space electrons pass through mass objects easily; some frequencies of the wave may bounce off, be absorbed, or pass through with only a comparatively small speed change, like visible light waves through glass. Visible light waves can travel many miles through glass fiber with little loss.

The primary difference in the wave carried by the free space electron versus sound is when sound strikes a particle of matter it must start the object to vibrating in order to carry the sound wave. Waves carried by the free space particles may pass through solid mass objects still carried by the particles. This causes many and varied actions, reactions and phenomena not possible with sound.

It is interesting how the light waves are propagated. The motion of matter propagates them all. If my memory is good, a famous scientist once said in one of his books that it is hard to believe that the motion of matter propagate EMWs. My personal view is that there is no other way to generate free space electron waves.

Enough of this. Let us move on. Inertia should be a good one. Let us try to make some sense out of inertia. There is a strong probability that I should leave it alone. No one else seems to be concerned as too why it exists or why it's very difficult to distinguish the difference between gravity and inertia. This is going to be difficult to put into understandable words.

Ok, *inertia of rest* will be our subject. Inertia of rest could be described as the resistance matter experiences when it is accelerated through space. It has already been stated that all forms of matter are under pressure on all sides by the constant bombardment of free space electrons, which, for the most part, pass on through, but a portion of them will strike something solid and bounce back in the direction from which they came. When soled matter is accelerated in space filled with free space electrons, a greater number of particles will rebound in the direction the object is being accelerated then any other direction. This of course, requires effort, or the consumption of energy. This effort will continue as long as the object accelerates. This effort or resistance to acceleration is called inertia of rest, or it could be called the resistance mass objects experience when they accelerate.

We will move on to *Inertia of motion.* When the acceleration is stopped and the mass object continues on at a constant velocity through the free space electrons, no additional effect to make it move will be needed. It will continue on until some force acts on it again. This is true because the rebounding, free space electrons that are rebounding off the mass object will strike other particles that are in front of the object and give up their forward kinetic energy to the incoming particles. This lessens the particle pressure on the forward side of the moving piece of matter. Consequently, the pressure will be the same on the front side of the moving mass object as it is on the trailing side. It will continue on forever through space, filled with free space electrons. This is called inertia of motion.

You may be having a little trouble believing that a particle can slam into a piece of matter and rebound back where it came from without slowing the mass object down a little. Part of your doubt of the credibility of this statement may come from a law of physics that states "for every action there is an equal and opposite reaction". Well, I am having a little trouble also. You might begin to get some idea as to its truth if you think of the particles colliding with something many thousands of times larger and heavier than itself, like the so called nucleus of an atom. We might also consider that the free space electrons rebound away at about the same speed as it came in at. This means that they lost no energy in the collision. If the particle expended no energy on the, solid matter it could not slow it down unless we create energy.

Even in this new way of thinking of our Universe we do not create energy out of nothing. So we must believe that the solid does not slow down as it travels through space, constantly receiving head-on collisions with billions of high-speed particles. An explanation of how this strange phenomenon might be accomplished must be given here if this one particle theory of the Universe has a chance of being believed. I can envision how this is done, but, if you have ever envisioned a difficult process, you may understand my problems in putting it in words. It would be helpful if you remember that many of the particles pass on through the piece of matter freely, with little effect on anything, for only a small part of the object is solid matter.

Ok, this is the way it's done. The piece of matter is traveling through space, which is densely populated with energy particles, free space electrons. We will build a special imaginary situation where one particle collides with one nucleus of an atom in the piece of matter. The particle then rebound away from the nucleus with approximately the same speed it approached it. The nucleus that was hit will slow down and be forced back into the mass object a minute distance where the particle pressure within the mass object returns it to its original position. This process will cause the whole object to slow down. However, the nucleus, or molecule, that was hit, in its process of stopping in its original position, will allow its own inertia to pull the mass object back to its original speed. This word pull does not fit the rest of the article very well for there are no attracting forces in this Universe. So a little explanation of this is necessary. In solid matter, the molecules occupy a definite space in the object. The molecules are held in this space by the particle pressure of the free space electrons in the mass object. This space is where the particle pressure is the same on all sides of the molecule. Move it out of equilibrium, and it will return to its original balanced position.

It may be difficult for you to envision how a molecule can be held in a near stationary position in a solid without any rigid restraints. It happens in all solid matter. You already know that all the voids between and inside of molecules are filled with free space electrons. These electrons are moving at velocities greater than the speed of the light wave and are capable of doing large quantities of work. When a molecule in a solid tries to move close to a neighboring molecule, the ping-pong action of the free space electrons between the molecules tries to push it back to its original position. A molecule may be free to move in any direction in a solid, but it is always forced back to its point of equilibrium.

It may be hard to visualize what we are calling ping-pong action, so we will try to make it happen on a ping-pong table. Bounce the ball on the table with a paddle, and see that the closer to the table you move the paddle, the more often the ball hits the paddle. It also strikes the paddle with greater inertial force "speed". The increase in frequency and greater force adds up to greater force or pressure. A greater pressure occurs whether we are talking about ping-pong balls bouncing between a table and paddle or free space electrons bouncing between molecules. They both represent a similar reaction and this reaction is how molecules are held together. Scientists that work with quantum mechanics would understand the ping-pong actions. They know the electrons, when confined to a small space, will become franticly wild, with unpredictable velocities and bouncing off of everything there is to collide with.

This ping-pong thing could be called a vibration instead of ping-pong action; vibrations in which the free space electrons reach a velocity greater than the speed of the light wave. The particles bouncing off of one molecule may have some collisions among themselves before this action reaches another molecule. The effect is still the same and will keep the molecules in a near stationary position. Of course, this reaction of the molecules trying to stay in equilibrium will be very rapid, so it could be said that the molecules are constantly vibrating. The vibration of the molecules is very rapid, but it is nowhere near the greater than light speed vibration of the free space electrons.

We have wandered a little from the subject of inertia, but the structure of matter must be understood to understand inertia. The similarity between inertia and gravity should also be understood, for they really are the same forces. To personally feel the effect of inertia and gravity, they both feel the same. In gravity, the free space particle pressure pushes you towards the earth where the particle pressure is less. The inertia effect is felt when you accelerate through space filled with particles. The particles you are passing through push on you, and the effect that is experienced is exactly the same as the one experienced in a gravitational field. The inertial force carried by the free space electron is the same as the force called gravity.

The above explanation of inertia is necessary, even though it is a natural physical property of matter to continue in motion and, when at rest, to resist acceleration. The reason for the explanation is that all space that solid matter moves through is crowded with free space electrons. It does not help that in this Universe, there is no friction. There is still the mechanical action of pushing away the particles that do not pass on through the mass object. The above explanation would not be necessary if the free space electrons did not exist. The fact is that all of nature's physical phenomena must fit together well with no special set of rules for different phenomena.

It's time to move on again. I feel it is time to look at where we have been. We have covered a lot of ground together. *Gravity, electrical,* the *strong* and *week forces* have all been tied together. The inertial force carried by the free space electron propagates all the four forces. A more accurate statement is that all four forces are only different facets or phases of our free space electron. One of the most sought-after feats in modern physics has been to show the interrelationship between the four forces. It has been accomplished here by uniting all the four forces of nature into one force. The free space electron carries the only force in nature, **inertia**!

In addition to this, we have shown that mass and energy just represent a mechanical change of state of the free space electrons. The solid form of matter, protons and neutrons, are only a cluster of free space electrons in a static state at rest whereas the energy form of matter is matter mechanically broken up into individual particles. Matter becomes the highly mechanical kinetic energy of free space electrons.

It is time for an observation again. The observation is that it should be becoming obvious that the true nature of the Universe, with all the phenomena and forces that govern it, cannot be understood by studying only a small segment of it at a time. All the physical phenomena are interrelated and live together in an orderly manner. This means that all the forces in the Universe must be studied all at one time. For scientists to specialize and study only a tiny segment of the broad spectrum of events in the Universe, it's unlikely that a complete understanding of nature will develop very rapidly.

Matter and Energy is so dependent on the electron existing as the free space electron, that I feel one more attempt should be made to make them more visible. *Molecular Motion* will be the subject. The assumption is made that you know the *Brownian Motion* states that microscopic particles of various substances in a fluid are held in suspension by the movement of the molecules in the fluid. My problem is what keeps the molecules in motion. We will build a problem for the molecules and see if they can stay in motion without the free space electrons to help them. A cylinder full of air molecules will be user for the demonstration.

First, we will remove the air from the cylinder to create a vacuum. Next, microscopic dust or smoke particles will be added to the cylinder. The cylinder is now full of particles that seem to be suspended in space. However, if the cylinder is left undisturbed, the dust particles will settle out to the bottom of the cylinder. The force of inertial gravity has forced them to the bottom of the cylinder. Now, we fill the cylinder full of air and again, the dust particles seem to be suspended in space. This time, they do not settle to the bottom. Instead, they dart around in a wild dance that will continue forever. It is the high-speed collisions with air molecules that keep them dancing.

The physics books state that the air molecules in a cylinder collide several billion times a second and will remain in motion forever with no outside force to maintain them. The reason, according to the physics books is that there is nothing to make them stop. The writer of this paper cannot agree with this. The air molecules do work and expend energy when they lift the dust particles from the bottom of the cylinder, and they will continue to lift them every time they are forced back down by inertial gravity. This means that the molecules will lose energy and eventually settle out if they do not receive some form of energy from an outside source. They do not settle out. It's no surprise to you or me what supplies the energy from outside the cylinder. The free space electrons pass through the cylinder walls with such ease they hardly notice they are there. They carry tremendous kinetic energy with them and can bat the air molecules around forever. The air molecules collide with themselves billions of times a second.

The free space electrons are thousands of times smaller than an air molecule, so the free space electrons probably collide about 3,000,000,000,000 times a second. That should be sufficient energy to keep the air molecules, dancing forever.

The above scenario is a little over-simplified. The vibrating free space electrons in the cylinder are small, and there are millions of them striking each air molecule from all sides. This would not make the air molecules move in any direction. To completely understand how the air molecules are kept in motion, we need to know how molecules in a gas function. The molecules in all gasses behave in a manner similar to the way they do in a solid. In a soled, molecules are held nearly stationary at a point of equilibrium by the free space electrons. In a gas, the molecules are too far apart to be held stationary. The free space electrons hammer away at them equally from all sides. This means that the gas molecules are not moved by the free space electrons until two molecules move close enough together that the ping-pong action of the free space electrons between the two molecules will cause them to repel each other violently. This is not like two molecules compressing a spring on there near collision. The free space electrons carry kinetic energy in their greater-than-light speed collisions. When a large quantity of these particles start their ping-pong action between the molecules, their inertial energy can be transferred to the molecules. This will cause quite a reaction with their greater-than-light speed collision. The result is that the air molecules will do their wild dance as long as there are other molecules to react with. We now have the air molecules powered, so they have ample energy to keep the dust particles dancing forever.

Back to free space electrons colliding so many times a second. If the electrons do collide at such a rapid frequency, they must be extremely close together. These greater-than-light-speed particles must be so close together that they are nearly a solid. There is a belief within a large segment of our scientific community that there is a considerable quantity of mass in the Universe that is invisible. It is sometimes referred to as **dark** matter or **invisible, elusive gray** matter. It has been stated many times in this paper that the free space electrons fill all space. The **electrons** are the **invisible, elusive gray** matter. It is not so elusive. It's everywhere. You can feel it as it pushes your body towards the earth. You can see it as it expends energy doing all the work that is done in the Universe. If ever the proverbial phrase "you can't see the forest for the trees" were true, it certainly is true here. The free space electrons can be called the invisible, elusive gray matter, or call the ultimate mass particle an electron, call it the particle in the particle wave theory of electromagnetic waves, call it the medium the electromagnetic wave is carried in, call it the presser with in the Universe caused by inertial gravity as it sweeps distant galaxies on the fringes of the Universe farther apart with accelerating speed, call it the particles that all mass objects consist of. Whatever you call it, electron, dark matter or ultimate mass particle, it certainly is not invisible. It is a monster, and I have a hold of it by the tail where it is showing the power it has with such wild gyrations that it will probably shred this paper before any one has a chance to read it.

I got a little carried away here, but I do hope this scenario has brought you a little closer to the possibility of believing that the free space electron is capable of doing what the writer of this book thinks it does.

It should not be too difficult for us to believe the free space electron carries large amount of inertial energy. If you are thinking, you already believe it! Probably you do not believe that last sentence, so we will do a little mental exercise here and see if I can bring you around. Think of a cylinder, such as a scuba diving tank, filled with air at 3000 psi. This means that every sq. in. of the cylinder area has 3,000 pounds of pressure on it. There would be over one million pounds of total inertial pressure exerted on the cylinder walls by the free space electrons. You have to believe that electrons are capable of exerting large amounts of inertial pressure unless you think the pressure on the walls of the cylinder to be nuclear reactions. No one believes the nucleus of the atoms themselves bang into the walls of the cylinder, so there is nothing else but the ping-pong action of the inertia carrying free space electrons to do the job. The free space electron does exist; it carries large amounts of inertial energy.

A short mathematical description of the **free space electron** may help you to believe that they are capable of possessing the tremendous energy needed to supply the whole Universe with all its energy needs! You may also find this description interesting. To start with, a cubic inch of space packed full of free space electrons would weigh billions of tons. This is also the weight of a cubic inch of matter in a neutron star. Sense neutron stars are made of neutrons and neutrons are composed entirely of electros at rest; this makes billions of tons the weight of a cubic inch of space packed full of free space electrons. The free space electrons throughout all space are packed closely together. Just pick a number; say there is one cubic inch of particles in every cubic foot of all apparently empty space. This means that a cubic foot of apparently empty space, filled with only this small percent of particles, would weigh billions of tons. Their velocities between collisions among themselves are tremendous. We might get some idea as to their real velocity if we compared it with the known speed of air molecules. Molecules of air reach speeds of .3 miles per second and carry sound waves at .2 miles per second. The air molecules are also known to collide a billion times a second. This means that air molecules travel 1.5 times faster than the wave they carry. The velocity of the light wave is 186,283 miles per sec. Using the same ratio, the free space electrons carrying the light wave would travel 1.5 times the speed of the light wave. (There is no experimental proof of an electron having a velocity greater then the speed of the light wave).

This means that the **velocity of the free space electrons**, between collisions, would be enough to orbit earth 11 times a second and collide many billions of times a second. Now, combine this velocity with the tremendous weight of these tiny dense particles times the number of times they collide a second and you will come up with an energy level that is unimaginable. This is the **mother** of all forms of energy. It is the live fluid energy that powers our **entire** Universe. I hope you enjoy this description of our particle as much as I do.

In the above paragraph, we talked about the particles in empty space being packed closely together and at the same time, occupying only a small present of space. If every cubic foot of apparently empty space has all of these particles within it compressed into a solid ball we would find that the balls would be more than a foot apart. However, if we break up these balls into electron-sized particles, the individual particles will be very close together, probably about the diameter of the old Bahr atom. It is difficult for me to envision distances closer than one thousandth of an inch. I can see, understand, and have worked with these measurements. These vibrating particles are much closed together than this; so far as I am concerned, they are very similar to a solid.

It is a solid only in the sense that the particles are very close together. Actually, it is a very lively fluid, for the particles try to fill all available space. Their energy level, which consists of inertial velocity, is all that keeps them apart as they collide with themselves. If the energy "velocity" of the particles were lost, they would become solid matter with no potential energy. This apparently happens in areas of space such as black holes, neutron stars, and the *proton* and neutron. We could say these collapsed particles are just another phase, or state, of matter.

There is a statement that is often made in this book that should be explained. The statement is that free space electrons move through a wire, or any mass object, or even move through free space. This is not completely true. It is their *effect* that moves through free space, or matter, with the speed of light. This is accomplished by forming wave actions of high and low densities in the free space electrons. The free space electrons may not move around in space as much as you think they would. They may be a better stationary reference point in the Universe than a large piece of matter. All you have to do is pick out a red one and use it for the reference point. Enough of that.

There is another interesting function the free space electron posses; it **governs** the speed of the light wave! This is true for the free space electrons carry these waves and any variation in their velocity will cause a variation in the speed of the light wave. This is as long as the means velocity of the free space electron remains constant; the velocity of the wave they carry remains constant "186,383 mi. per sec. in a vacuum". This of course predicts the speed of the light wave will remain **constant** and not be effected by the **motion** of the **source** of the light wave or the **motion** of the **observer!** The velocity of the free space electron varies a small amount when they are **within** or **near** the various forms of matter. Near large pieces of matter, such as our planet earth, the free space electrons lose some of their linear velocity. How this velocity was lost was discussed earlier in this book under the heading of gravity. This slowing down of the free space electrons, cause the so-called constant speed of the light wave to also slow down a proportional amount! I hope you like the way the speed of the light wave is controlled by our free space particles.

This may be a good time to discuss and review the true nature of matter in our Universe. The Universe consists of *matter* and *energy*. In this Universe, matter and energy are one of the same. **Energy is only the inertial motion of matter!** There is no other form of energy in our Universe other than the **kinetic energy of matter.** Matter is generally thought of as something that occupies space and can be perceived by the senses. A large portion of matter in the Universe is not so tangible. It is not easily detected by our senses. We also think of the composition of all matter as existing in three states—gases, liquids, and solids. **Matter exists in five stated in this Universe.** These five states of matter are the building blocks of the Universe, complete with all its energy and forces.

The five States of Matter

The *first* and primary state, or we could call it the first phase of existence that matter exhibits, is the free space electron which is the *Ultimate Mass Particle* it's self. It is the primary building block of the Universe. Its weight and velocity were discussed above. It is a solid particle of matter that occupies all of apparently empty space. It travels at tremendous velocities between collisions among its own kind throughout all space. The velocity of the particle carries nearly all the energy in the Universe. It, the **free space electron,** is the **first state** of matter in our Universe.

The *second* phase or state these particles of matter pass through is where the free space electrons lose energy and collapse into solid matter. The solid matter we are concerned with are protons and neutrons, which are the building blocks of the atom.

It might be interesting to know what we mean by lose energy. This energy loss is the loss of inertial velocity the free space electrons possessed before they formed protons and neutrons. Their average velocity between collisions among themselves is 1.5 times the speed of the light wave, or more correctly stated, the speed of all waves in the so-called electromagnetic spectrum. Therefore, the protons and neutrons have very little inertial energy when comparing them with the large amount of energy the free space electron possess. This is like forming a proton by removing 1837 free space electrons that had no relative motion at the time they were removed. This leaves all of the electrons that reside in free space with a little more average energy than they originally had. Stated differently, the number of electrons in free space has decreased and their average velocity has increased. The total energy level of the free space electrons has not changed.

This second state of matter which consists of protons and neutrons, could be called the element hydrogen, which can consist of no more then a lone proton. The process of forming hydrogen from free space electrons is where the free space electrons are forced together and lose their inertial energy and form a proton. This proton is probably formed in the harsh environment of a supernova explosion. This is where tremendous pressures exist that are capable of forcing the electrons together. This process of forming hydrogen out of free space electrons can be reversed; all we have to do is heat the hydrogen atom "probably by an atomic reaction" to a very high temperature or heat. Another term for heat is molecular motion. The molecular motion causes enough mechanical motion within the hydrogen atoms to make them fly apart and form new atoms of helium. Some of the neutrons do not survive the transition of becoming helium atoms. They are smashed and broken up into the free space electrons, since all matter is made up of free space electrons. These smashed neutrons cause the surrounding area of space to be greatly overpopulated with electrons. Here, the high speed pounding action of the fee space electrons cause the static electrons of the mass fragments to disperse, in an expansive action. They gave each electron that was involved in the reaction a push, and they became what we think of as free space electrons, "**energy**". They moved for more uniformity, like that of the other free space electrons. This, of course, is the violent expansive action called thermonuclear reaction, Einstein's mass energy theory $E=MC^2$. You will notice that matter was not destroyed to create all of this energy. It was only matter changing state, from the second state to the first state. Matter is indestructible. All matter can do is change state.

Back to the states of matter where we will work with the *third state.* The third phase that this matter passed through is a *gaseous state.* This was discussed in some detail, under the heading of the Brownian Motion so it should not be necessary to go through it again. We also know that protons and neutrons are composed of electrons at rest where they are the building blocks of the atom. These atoms and molecules make up the solid portions of our gaseous matter. These gaseous molecules are not very close together, so they fly around freely until they come close to another molecule. Here the ping-pong action of the free space electrons between the molecules forces the molecules of gas to fly off in some new direction. The gas molecules are more or less free sprites and tend to fill all space available to them. Their only restriction is when they make close contact with another gas molecule or a molecule in some solid matter object. That illustrates the third state that matter exists as.

The *fourth state* of matter is a liquid. It is very similar to that of gaseous matter. The molecules are free to move anywhere in the liquid until they come too close to another molecule where the ping-pong action of the ever-present free space electrons will violently force them apart. They do not seem to be able to find a point of equilibrium. Under some circumstances, a few of the molecules may even fly out into space as a gas. But, for the most part, the molecules do not reach enough speed to leave the liquid. It also means that the movement of the molecules in the liquid is restricted enough so that the molecules remain a liquid and do not try to fill all available spaces. That is the way molecules function as a liquid.

The *fifth state* of matter is a solid. We worked with the structure of matter in a solid under the heading of *inertia,* so you should be somewhat familiar with it. As you remember, the molecules are not free to move about in a solid. They are held at a near-stationary point in the solid. This is true, for, at this point, they are at **equilibrium** with the pressure around them. This point of equilibrium changes when solid matter travels at high velocity through free space filled with free space electrons. This motion causes a large quantity of particles to flow through the solid matter where it forces the molecules, and the point of equilibrium, to move closer together in the direction of motion. Moving the point of equilibrium, and molecules closer together **predicts** that solid matter will become **shorter**. The pressure within the solid is caused by the free space electrons, which fill all the apparently empty space, within a solid. If a molecule moves in any direction, it will come too close to another molecule where the ping-pong action of the free space electrons will force it back to its original positions. This process is continuous; so you might say the molecules in a solid are constantly vibrating. This vibration or "molecular motion" can be increased or decreased by placing the piece of solid matter in different environments. The motion of the molecules is very important for us to understand. The physics books would call this motion heat. The author of Matter and Energy does not like words like heat for they muddle up your thinking, so we will leave it at molecular motion.

Back to solid matter. On the outside of solid matter the perimeters are held rigidly stationary by the strong free space electron force. This is the fifth state or form matter takes in this Universe. It should always be remembered that the free space electrons make up the composition of all the forms of matter, as well as propagate all the forces in nature. With the use of these *five states of matter* and **inertia,** **all** of nature's mysterious wonders and forces can be understood.

Solid matter is not always stable. An unstable atom is one that is most likely to large for the free space electrons to hold it together properly. The atom is likely to split and form two atoms of a lighter element. During this harsh transition period, some of the neutrons are likely to be hammered hard enough "by fast-moving nuclear parts and free space electrons" to make them break up into their basic particles, electrons. These clusters of electrons will continue to be hammered by the free space particles until they get moving and dispersed fairly evenly in space. We could say that the immediate space is expanding. Along with this expansion of space, some large particles of matter that did not break up into electrons may fly bout violently. All of this violent outward movement of particles is what is called atomic reaction, or **nuclear fission**. All of this energy release came from the free space electrons as they slammed into the **smashed** nuclear parts. It could also be said this energy came from a change of state. The **fifth state** of matter changing into the **first state** of matter, which is the same as a solid changing into the greater-than-light-speed free space electrons.

We have been doing a lot of work on the structure of mater; however, something of importance has been lift out. This is the molecule. How it is formed and held together is necessary for a complete understanding of solid mater. Molecules, of course, are small pieces of matter made up of atoms. The atoms in a molecule are held together very similarly to the way molecules are bonded together in solid matter. The molecules finding a point of equilibrium is the bonding force that holds solid matter together. The atoms in a molecule are held together the same way. They, too, find a point of equilibrium. A lone atom is constantly being bombarded by greater-then-light-speed free space electrons. This causes a very high particle pressure to completely surround the atom. When two or more atoms are forced together, they fall into the low-pressure area near the other atom or atoms. The low particle pressure area is created by the shielding effect of the atoms themselves. Free space electrons cannot penetrate the atom. Consequently, there are no particles coming out of the portions of the atoms that are facing each other. This makes fewer particles "that possess large amounts of linear motion" between the two atoms than there are on the other surfaces of the atoms. I like to call this a **shadow effect** in a sea of very energetic inertia carrying particles. Therefore the free space particles will force the two atoms together. That is, they are forced together up to the point that the ping-pong particle pressure between the two atoms becomes greater than the particle force that is forcing them together. This means the atoms in the molecule are in constant motion "heat". So the atoms within a molecule vibrate back and forth trying to stay at a point of equilibrium. This is the way atoms form molecules. There are many reasons why atoms will **not** form molecules. Probably all the reasons are simply known properties of the different atoms, such as size, configuration, inertial mass, spin, size of shadow, velocity, etc. One example would be if two atoms came together at too great a speed, the ping-pong action between the atoms would force them back out of the shadow and on out into free space. They would not bond with each other.

In the above paragraph the shadow effect was mentioned. It should be of interest to note that this shadow effect is the same shadow effect that was a major factor in propagating inertial gravity!

In classical physics, it is thought that solid matter absorbs heat in two different ways. One is to increase the motion within the molecule "heat", and the other is to increase the motion of the molecules in a solid, which also is heat. The above theories predict that this is true.

There is something of importance that may not have been noticed in the above description of the nature of matter in this Universe; the **atom was redesigned.** It is no longer large like it was. The diameter is **only** the diameter of the nucleus. The neat rings of electrons circling the nucleus of the Bahr atom do **not exist**. The neutrons and protons in the so-called nucleus of the atom are all there is to the atom. The so-called nucleus of the atom is taken good care of by the free space electrons without organized orbiting rings of electrons orbiting the nucleus of the atom.

If you read this article, you know how the atoms or molecules are held more or less rigidly at a point of equilibrium in a solid. The atoms or molecules are able to react at a distance with the neighboring atoms or molecules without the aid of organized orbiting particles around the atoms. I like to call the distance at which atoms are capable of reacting with other atoms the **reaction zone.** The physics books call it the distance the outermost ring of electrons is from the nucleus of the atom. The diameter of the atom is the diameter of the outer ring of electrons, whereas in Matter and Energy the diameter of the atom is **only** that of the so-called nucleus. The *reaction zone* is where the free space electrons create a ping-pong repelling action between **atoms or molecules.**

The atoms that most of us envision do not work. To try to put this in proper perspective, try to envision the nucleus of the hydrogen atom "one proton" to be one mile in diameter, "big". This would make the lone electron that orbits the nucleus 5 feet in diameter and 500 miles away from mother nucleus. There is no logical way this five-foot ball of matter

500 miles away from mother nucleus is going to establish the perimeter of the atom and do all the things it supposedly does. Stated differently this means the perimeter of the old atom has a surface area of 785,398 square miles. This one five foot ball 500 miles away must constantly patrol and try to keep all particles of matter, large or small, from colliding with the so-called nucleus of the atom. This is preposterous and totally beyond reality. Matter and Energy calls this region of space between the so-called nucleus and the outer diameter of the atom, the reaction zone. Conversely the new atom would have a hole Universe of five-foot balls completely surrounding the so-called nucleus, colliding with everything there is to collide with in the reaction zone. They would travel at velocities 1.5 times the speed of the light wave, and be capable of establishing a boundary for the atom.

Another point that you may have missed is a description of the **weak** and **strong** nuclear forces. The weak nuclear force could be described as the force between protons that tries to force them apart. The closer the protons approach each other, the greater is the repelling force. When they are very close together or almost touching, it is a very strong force. This is true for as you remember, all of space, as well as all space around a lone proton, is filed with free space electrons, ping-ponging back and forth between the protons. This very strong hammering-repelling action between the protons is what we call the weak nuclear force.

The strong nuclear force will arise when two or more protons or neutrons touch each other. The physics books states, the strong nuclear force mysteriously takes over at this time, and the repelling force between protons stops functioning. The protons and neutrons will now attract each other with an extremely strong force). **This is not true.** There are no attracting forces in the Universe. The protons and neutrons within the so-called nucleus are forced together by the powerful free space electron force outside the nucleus. This is true for, at the point where the protons and neutrons touch each other in the so-called nucleus of the atom; there is no particle pressure. There is no particle pressure at this point because the free space electrons can no longer squeeze between the protons and neutrons to do their ping-pong thing. This allows the free space electrons to force all the nuclear parts together. This is how the **strong** nuclear force **really** functions.

The strong and weak nuclear forces do not really exist! The phenomena called the strong and weak forces are nothing more than the free space electrons reacting naturally exerting their normal inertial force in and near the atom. In fact, there is only one true basic force in the Universe! This force is *Inertia*! The free space electrons, as well as all matter possess Inertia. The **inertial** effect of the free space electrons, and the various conglomerations of matter composed entirely of electrons, is the only **true force** in this Universe!

There has been great difficulty in understanding the mechanics of the four forces: *gravity, electrical,* the *strong* and *weak* nuclear forces. The reason there is a poor understanding of these forces is that they do not really exist. Man created the four forces in an **unsuccessful** attempt to understand the observed phenomena of nature's forces at work. Take gravity for an example. Gravity is thought to be some kind of a magical attracting force possessed by all forms of matter. This force tries to pull all mass objects together. **Not true**. We are trying to create a force where none exists. All matter is simply pushed together when it is hit by the inertial force carried by the high velocity of our free space electrons. How this is done was discussed earlier in this article. It was known in the time of Isaac Newton that you couldn't tell the difference between inertial and gravitational effects. There cannot be two different forces that produce identical effect. I personally think that Newton should have been criticized for not trying to give an explanation for this in his work with gravity.

In Matter and Energy, gravity loses and the inertial force of matter takes over the job that the imaginary force, gravity, was believed to have done! In reading this article, you should have noticed that inertial forces of matter have taken over the functions of **all the forces!** The entire Universe is just one big beautiful inertial system. With the use of **inertia and the free space electrons, along with the natural shielding or shadow effect that all matter possesses, all of nature's wonders can be understood.** This is the *foundation* of the "*everything" theory* that Einstein wanted so badly. Einstein wanted this theory so much that he spent the later years of his life trying to discover it. Apparently he believed there could be such a theory. Applying the "everything" theory will give us a **complete** understanding of all-natural phenomena and forces.

I feel that it is time to do something different with this book. Something different like build a new theory on how the Universe began, using the "everything" theory for the foundation to build on. Hopefully this can be done and stay within the boundaries of **reality.** An interesting hypothesis on how the Universe we now live in began could be like this.

Assume there was a primordial Universe consisted of no solid matter such as protons and neutrons. There was only a tiny dot of matter the size and mass of the electron. There were many of these dots scattered throughout the primordial Universe. They were spaced apart at a means distance equal too less then the diameter of the Bahr atom. Now imagine two monsters inertial shock waves of these particles colliding. Two primordial Universes colliding could have created these waves. As these particles in the two Universes crunched together and through each other, many large clusters of particles were forced together. They stayed together due to inertial particle pressure that fills all free space and surrounds all solid matter. Clusters of particles the size and mass of protons and neutrons were the easiest to hold together. They were the most stable clusters of particles enabling them to still be with us today. As the Universes continued to over run each other even grater quantities of solid matter were created from these tiny particles. Every free space electron that became solid matter gave up all of its inertial energy to in-coming free space particles. This would give the particles "in the vicinity of the collision" enough inertial velocity "energy" to start a reversal of the two shock waves. This gave the individual free space particles more velocity and inertial energy then they had when the two waves collided. This means that more matter was being precipitated "change of state" as the Universe began expanding very fast. Enough matter was formed to build the Universe we have today. As time passed the two Universes continued to expand as one. In our present-day Universe the velocity and pressure of the free space particles has decreased, however their velocity is still 279,000 miles a second *between collisions among themselves* and the inertial pressure is still tremendous. The great quantities of solid matter have broken up into protons neutrons and free space electrons forming our present-day atoms of matter. It should become apparent that excepting the existence of the free space electron all physical phenomena can be understood!

The most popular theory about how the Universe began goes far beyond any semblance of reality. In this theory the Universe began from one tiny little particle of matter and this particle grew into a complete Universe. This is not possible while staying within reality. What can be done is build the Universe out of **many** small particles of matter, providing they contain enough matter to do the job. This is staying within reality which is a very important thing to do.

It would now be appropriate to answer the question "will the Universe expand forever"? In answering this question we will start by examining the most distance galaxies. These distance galaxies, on the fringes of the Universe, accelerating away into never never land seams to indicate the Universe will expand forever. With all the particle pressure of the free space electrons pushing the galaxies farther apart and on out into space it appears impossible to stop this action. The very fact that the entire Universe is expanding will **stop** the expansion and start it to reverse its direction! Start contracting.

This is the way the *impossible* is done. As the Universe expands everything becomes farther apart. In the beginning of our present day Universe the free space particles were very close together where they collided with each other at tremendous velocities and pressures. Their velocities continued to increase until the expanding Universe caused these particles to become so far apart that they could no longer generate enough pressure to precipitate solid matter. As the Universe continued to expand the inertial pressure of the particles could not create enough pressure to hold matter together. The so-called nucleus of the atoms, on the fringes of the Universe began to brake-up in to protons and neutrons. As the expansion continued the free space electrons became spaced even farther apart where they could not even generate enough inertial pressure to hold protons or neutrons together. Consequently they broke-up into their basic building blocks, electrons. Huge quantities of these electrons began to form. At this point the free space electrons crashed into these particles causing them to disperse evenly into free space like all the other space particles. This caused an explosive increase of inertial pressure which pushed on the distant galaxies causing them to accelerate on out into space. Earlier in this article under the subject of nuclear reactions it was described how and why the liberation of free particles caused large increase of inertial pressure. As time passed even greater quantities of free particles formed from the solid matter that still existed in the distance galaxies. These newly formed particles absorbed huge amounts of inertial energy from the free space electrons. There was so much inertial energy taken from the free space electrons that it caused inertial particle pressure to drop very low in and near the fringes of the Universe. This allowed inertial particle pressure outside the Universe to start it to contract; reverse its direction. As the Universe continued to shrink, more solid matter decays into particles, absorbing inertial energy from free space electrons. Eventually all or nearly all solid matter is reduced into electron. As all these particles move toward the center, collide and overrun each other the **entire process starts all over again.** The Universe starts to rebuild its self.

An interesting observation can be drawn from the material above that describes the various phenomena that takes place on the fringes of the Universe. This observation is that all actions and reactions on the edge of the Universe are almost like having an enclosure built around our Universe. That's **wild** isn't it!

One more point needs to be made. This is with all the commotion going on near the edge of the Universe it would seem logical to think of this place as being extremely hot. This is not true. It is colder then the Bose-Einstein condensate where all molecular motion stops. The free space electron which normally has an average velocity of 279,000 miles a second at 3 degrees k. It is so cold, on and near the edge of the Universe, the electron as well as all matter has lost most of their motion. I like to envision the edge of the Universe as being a place where the motion of mater has become very sluggish.

There is a prediction that can be drawn from the above paragraph. The prediction is the Bose-Einstein condensate will have no viscosity. The reason being, Due to the lack of free space electrons pressure all the matter in the condensate, protons and neutrons, have decayed back into there basic building blocks, free space electrons. These electrons posses no relative motion. There is no type of construction material to build a vessel that the electrons will not run right on through. These electrons are the smallest and the most dense particle in the Universe. They easily run right on by the protons and neutrons in the construction materials the vessel would have to be made of. These no-velocity electrons are found in huge quantities on and near the edges of the Universe as it begins to shrink. If you expose a quantity of this stuff to the normal free pace electrons an extremely violent reaction would occur. This pressure increase would increase the speed of the shrinking process.

The theory of how the Universe began was a very hard theory to write. What you just read in the paragraphs above may not be precisely how everything works however it does stay within the laws of physics and reality. This is more then present theories can say. It also reinforces the theory that free space electron exist, creating pressure differences that precipitate **all** physical phenomena. It should also stimulate your thinking, let you think of the Universe a little differently then you have been. Hope you liked it as much as I do. Its time to move along again.

One more point on the nature of matter should be made. Sub atomic particles, or fragments of nuclear particles will be the subject. By sub atomic particles we are referring to particles of matter that are not protons, neutrons or electrons. These particles that are mostly man made in a particle accelerator, are transition particles. After they are produced, there are few that survive more than a small part of a second. They are unstable, and most will decay back into free space electrons.

These particles do not really belong in the Universe. When they are built in an accelerator, or naturally, they are either too large or too small for them to be held together well by the free space electrons. This means they are not the size and weight of protons and neutrons. There is only one particle in the Universe that is entirely stable and indestructible. This particle is the free space electron itself.

The proponents of quantum mechanics believe this solid piece of matter, the electron, can disappear and turn into a wave. This makes the electron appear very unstable. The electron is **not** capable of turning into a wave and shortly it will be proven in this book. Protons are also very stable. Neutrons are fairly stable inside of an atom. All other known particles are transition particles and will decay back into protons, neutrons, or free space electrons.

Modern physics seems to be so anxious to understand the observed phenomena of these atom fragments that they are willing to ignore reality and the physical laws of science. Reality along with the laws of physics has been thrown to the wind and almost completely lost in the atomic-subatomic field. This reality thing is very disturbing. In Matter and Energy, the laws of physics have not been broken. The law of universal gravitation had to be bent a little. In this law the **arbitrary term** *attract* was removed and replaced with the term repel. The writer of Matter and Energy certainly hopes reality can be given back to the atomic-subatomic field.

The departure from reality is not a new thing. It probably began with early humans when they thought of gravity as being an attracting force. Attracting or pulling forces without a physical link to do the pulling is going beyond **reality!** I have also read about shepherds on a rocky hillside observing the rocks exerting a magical power of pulling on the nails in their shoes. These loadstones truly were an imaginary pulling force for there are no attracting forces in this Universe. The mechanics of attracting forces can **not** be explained. So as you can see science has been living beyond reality for a long time. This made it much easier for scientist to go beyond reality in the atomic-subatomic field. It "really" was not that large of a quantum leap.

At this time it seems to be appropriate to write a short note or observation about this book. Matter and Energy cannot be believed as long as *quantum mechanics* is thought to function far beyond the boundary of reality. Such things as a particle like the electron making a quantum leap and turning into a wave destroys the very foundation of this entire book. A solution to the problem would be to prove quantum mechanics could operate within reality. An attempt will be made to **do just that.** Suppose we take the strongest experimental proof "possibly the only known proof that matter can function in a manner completely beyond reality" and demonstrate how this can be done staying within the bounds of physical law "reality". In this experiment quantum mechanics has supposedly proved beyond any doubt that matter on the small scale, such as an electron, can and does function beyond the laws of physics. This experiment is of course the double slit experiment. As you probably know in this experiment electrons are fired at a target where the impact of the electrons will be recorded. Between the source of electrons and the target there is a partition, which has two holes, or slits cut in it. It is believed and supposedly proven that these particles of matter "electrons" stop being **particles** and become **waves.** These waves move from the source, through the slits and arrive at the detector where they suddenly change back into electrons. It is this experiment where quantum mechanics draws its **power**, proving that matter can instantaneously disappear and change into waves "quantum leap". These waves can also disappear and change back into a particle of matter, another quantum leap. Believing this allows quantum mechanics to ignore reality and the laws of physics.

Now if this same experiment can be done using the free space electrons and show these particles "electrons" have no duality, then quantum mechanics may be put back into the realm of reality and the free space electron may proven their existence. All of the phenomena described by quantum mechanics **can be done with the inertia energy carrying free space electron.**

It is really very simple how the free space electron function in the double slit experiment. This is how it is done. Matter and Energy has already shown how all free space in the Universe is full of free space electrons. This of course means that in this experiment all the space between the source of electrons and the detector screen is filled with very energetic inertia carrying free space electrons. These electrons are capable of carrying waves of all frequencies at 186,280 miles a second. So in this experiment all that has to be done is poke or accelerate a few electrons and they will send out a wave disturbance in the free space electrons that will pass on through the double slits and on to the electron detector screen. In this experiment the waves do not have to change back into electrons for the waves them selves are composed of free space electrons. If there are a large number of electrons accelerated at the source, there will be a large number of electrons absorbed at the detector. This means the free space electrons are given enough additional energy to force a large number of electrons to be absorbed by the electron detector screen. It is obvious in this experiment that they are electrons when they leave the source, they are still electrons when they become a wave or waves of free space electrons and they are still electrons when they are absorbed by the detector screen. This all can be done and stay within the bounds of reality!

There are the same results of this experiment when photons of light are used. This does not really need an explanation, for in Matter and Energy it is explained how the free space electrons carry the light wave and all the other frequencies of waves in the so-called electromagnetic spectrum. The photon does not exist. The energy in the light wave is carried from its source by the wave action in the free space electrons. It is not carried by the magical mythical photon. There is no quantum mechanics magic involved in this experiment. No quantum leap. With the use of the inertial energy carrying free space electron all of the unrealities of quantum mechanics can be explained and stay within the laws of physics! **This double slit experiment proves the *free space electron* does truly exist!**

It is strange that it has been so hard to correctly understand the wave action in the double slit experiment. This natural function of the free space electrons is found everywhere in the

Universe. It is this free space electron wave that carries all the energy from the sun to earth. This same wave carries the energy the earth radiates into the night sky. In nuclear power plants the very destructive gamma radiation are waves in the free space electron. **All** energy that travels through space at light speed is carried by waves in the free space electrons! This also includes electrical transmission lines where waves in the electrons that are already within the lines, carry all the energy. Nearly all the energy in the Universe is carried by waves in the free space electrons!

Another note of some importance. It should be remembered "when trying to understand our physical Universe" that free space is full of energy and matter that can be **harvested.** One example of harvesting **energy** from free space is the **fusion** and or **fission** of the atom. Another example of **producing matter** from free space would be when an electron and a positron collide in a particle accelerator. This collision produces a mass many times **greater** then the original particles. The additional mass came from the free space electrons. The most obvious way for matter or energy to be harvested from free space is through chemical reactions. All chemical reactions release energy when a nuclear part such as a neutron or an entire atom is smashed in the harsh environment of a chemical reaction. By smashed it is meant that a neutron is broken up into its basic building blocks, which are electrons. These electrons over populate the area around the so-called nucleus of the atom where they are bombard by the free apace electrons, which cause them to be dispersed evenly in free space. This violent outward moving action, of the electrons being dispersed evenly in space is the release of energy we are referring to. This is an ordinary nuclear reaction that will always harvest energy from the free space electrons. This could also be considered a change of state. The neutron changing back into free space electrons.

It is time for another small observation. This comes from a statement made by Einstein. The statement is that all particles of matter, large or small, in the entire Universe, attract each other. Free space has often been described as a giant vacuum that produces an attracting force that tries to pull all matter together. Contrary to this the writer of Matter and Energy believes that all particles of matter, large or small, in the entire Universe exert an inertial repelling force **when they collide.** One example of this is when the free space electrons collide with each other; there is a strong inertial repelling force caused by the electrons ping-ponging back and forth between themselves and **all matter they come in contact with.** These particles of matter cause a tremendous pressure throughout the entire Universe. This pressure becomes self-evident by observing the motion of the galaxies that reside on the fringes of the Universe. These distance galaxies are accelerated away from the Universe by the free space electron pressure and on out into the unknown. This is another **proof** that gravity is a repelling force, not an attracting force. **All physical phenomena in the Universe are created by a pressure difference!**

There are four points in Matter and Energy that I wish to stress. The first point is that our ultimate mass particle "the free space electron" does reside in **all** apparently empty space. The second point is that the new atom is composed of a cluster of electrons only and has no organize orbiting particles beyond the so-called nucleus. The third point is that all physical phenomena are created by pressure differences. The last point is that there is only one true force in nature, **the inertial force of matter.**

Hopefully Matter and Energy has made a few people believe the electron does exist in free space. If I did not make you believe in the existence of the free space electron, possibly this article will at least make you do a little thinking. The greats of the past may not have always pointed us on the right course. There is room for free thinking. Matter and Energy may have; just may have, found the key, to the lock, to the **Everything Theory!**

Know that we have the key, let us use it. I know where there are eleven doors that need to be unlocked. These are the doors to the eleven greatest unanswered questions of physics. The 11 Greatest Unanswered Questions of Physics came from Discover Magazine February 2002. The answers to these questions are revealed below.

Question number (1) What is dark matter?

Dark matter is the free space electron, which occupies all of apparently empty space. These electrons are spaced together at about the means distance of the diameter of the old Bahr atom where they account for most of the matter in our Universe.

Question number (2) What is dark energy?

Dark energy is the **inertial** energy possessed by the velocity of the free space electrons. Their inertial velocity between collisions is 1.5 times the speed of the light wave. They collide a tremendous number of times a second where they are constantly exchanging inertial energy with their greater than light speed collisions. These free space electrons are the dark energy in our Universe. To completely understand the first two questions it is necessary to read the book Matter and Energy, which is a one-particle theory of matter and energy.

Question number (3) How are heavy elements from iron to uranium made?

Heavy elements are formed in an environment of tremendous inertial pressure. This inertial pressure may be found in such places as supernova explosions where protons and neutrons reach extremely high speeds. When they collide with each other, in this environment, they are stuck. The bombarding pressure of the free space electrons will hold these heavy atoms together forever. That is forever until some mechanical force is exerted on them that is strong enough to knock them apart.

Question number (4) Do neutrinos have mass?

Their existence is questionable. It may be neutrinos, ultimate mass particles and free space electrons are all the same particle!

Question number (5) Where do ultra-high energy particles come from?

These most energetic particles that bombard earth are called cosmic rays. They consist of gamma rays and various bits of subatomic shrapnel. These tiny bits of matter are hurled out into space by the tremendous shock waves created by such sources of energy as a supernova explosion. These bits of matter are further accelerated to a very high-energy level as they ride the various frequencies of light waves in the free space electrons. They reach energy levels of near the speed of the light wave itself.

Question number (6) Is a new theory of light and matter needed to explain what happens at very high energies and temperature?

YES A new theory of light and matter is needed to gain a complete understanding of **_all_** physical phenomena. The new theory for light is that light is carried through space by a wave action in particles of matter called, free space electrons, **not by the photon. The photon does not exist.** These free space particles of matter have an average inertial velocity of 1.5 times the speed of the wave they carry and their average density in space is the same as the diameter of the Bahr atom. They posses and carry nearly all the energy in this Universe with their greater then light speed inertial velocities. Atoms of matter are thought to consist of protons and neutrons. Protons and neutrons are composed of the *Ultimate Mass Particles, called* **free space electrons.** A complete understanding of this new theory about light and matter can be achieved by reading the book Matter and Energy, which is a one-particle theory of matter, and energy.

Question number (7) Are there new states of matter at ultra-high temperatures and densities?

Under these extremely harsh conditions where particles of matter collide with very high speed, exerting tremendous inertial pressure on each other, matter can brake down into its basic building blocks. The basic building blocks that all matter is composed of are free space electrons. Protons and neutrons and all solid matter are composed of these particles. This ultra-high temperatures "motion of matter" and densities could also force these free space electrons together forming all kinds of wearied particles. Some large, such as super massive boson. There may be extremely large or small particles composed of the wreckage of protons and neutrons. So the answer to the question is **yes**, there could be new states of matter created in this harsh environment. What cannot be done is create a huge particle out of **<u>a</u> small particle** for this is going beyond reality. It is necessary to stay within reality in order to understand **all** physical phenomena. What can be done is create a large particle out of many small particles such as the wreckage of protons and neutrons, which are electrons. These large particles could also be built out of the free space electrons, which fill all apparently empty space in the Universe. To fully understand the answer to this question it is necessary to read Matter and Energy, which is a one-particle theory of matter and energy.

Question number (8) Are protons unstable?

Under the right environment all matter can be broken down into ultimate mass particles, which are free space electrons.

Question number (9) What is gravity?

Gravity is an **inertial repelling force**. Gravity is thought to be an attracting force which is not true. It is going beyond reality for pieces of matter to attract each other without a physical link between them to do the pulling. It is necessary to stay within the bounds of reality to understand our physical Universe. If the above statements are true it destroys all field theories that describe attracting forces. Einstein stated that **all** particles of matter, large or small, exert an attracting force throughout the entire Universe. Matter and Energy states that **all** particles of matter, large or small, exert an inertial **repelling** force throughout the entire Universe, **when they collide.** A simple example showing that gravity is a repelling force has been observed by studying distant galaxies on the fringes of the Universe as they accelerate away into never never land. The galaxies are pushed away by inertial pressure caused by the inertial velocity of the free space electrons. (Read Matter and Energy). Inertial forces are the only forms of energy in this Universe and the free space electron carries most of this inertial energy.

Question number (10) Are there additional dimensions.

No, Stay within reality.

Question number (11) How did the Universe begin?

There are various theories on how the Universe began. Most seem a little inadequate. However some new rules that govern how the Universe functions "found in the paper Matter and Energy" should be extremely helpful in answering this question. Some of the rules will be listed below. You will notice the term rules are used instead of hypocrisies, theories, or laws of physics. This was done so that "depending on the degree of acceptance" the readers of the rules can put their own name on them. Hopefully a disrespectful term is not selected. Matter and Energy should be a prerequisite for reading these rules for it will provide proofs and a better understanding.

Using the 11-**answered** questions, the 9 rules and the book Matter and Energy, it should be easy to form hypocrisies on how the Universe began. This can be done and stay within the boundaries of **reality.** An interesting hypothesis on how the Universe began "using the references above" could be like this. Assume the primordial Universe consisted of no solid matter, such as protons and neutrons. There was only a tiny dot of matter the size and mass of the electron. There were **many** of these dots scattered through out the Universe at a means distance apart equal to the diameter of the Bahr atom. Their velocities between collisions were 1.5 times the speed of the light wave. Now imagine two monsters inertial shock waves of these particles colliding. Two primordial Universes colliding could have created these waves. As these particles crunched together and through each other they were forced very close together creating tremendous inertial pressure. In this environment many large clusters of particles were forced together and stayed together due to the inertial particle pressure. Clusters of particles the size and mass of protons and neutrons were the easiest to hold together. They were the most stable clusters of particles enabling them to still be with us today. As the

Universes continued to over run each other even greater quantities of solid matter were created. Every free space particle that became solid matter gave up all of its inertial energy to incoming free space particles. This would give the free space particles in the vicinity of the collision enough inertial energy and velocity to try to start a reversal of the two shock waves. This gave the free space particles more velocity and inertial energy then they had when the two waves collided. This means that even greater quantities of solid matter was being precipitated "change of state" as the area between the two Universes began expanding very fast. Eventually the two Universes continued to expand as one. This is how the Universe began. It should be apparent that excepting the existence of the free space electron all physical phenomena can be understood.

There are nine useful rules that I wish to give you before ending this book. These rules were developed over a period of many years. Even though these rules and Matter-and-Energy grew over the same time period I have found them very helpful in developing good theories that describe nature's natural phenomena that are found in Matter and Energy. These rules are listed below where possibly you will find them interesting.

Rule number (1) All Physical phenomena must be explained staying within reality.

Rule number (2) Gravity is **an inertial repelling force**. It is going beyond reality for pieces of matter to attract each other without a physical link between them to do the pulling. If the above statement is true it destroys all field theories that describe attracting forces. Einstein stated that **all** particles of matter, large or small, exert an attracting force throughout the entire Universe. Matter and Energy states that **all** particles of matter, large or small, exert an inertial repelling force throughout the entire Universe, **when they collide.** One simple truth that gravity is a repelling force is by observing the distant galaxies on the fringes of the Universe as they accelerate away into never never land. They are pushed away by inertial energy possessed by the Free Space Electrons.

Rule number (3) Space is full of Free Space Electrons moving with an average velocity of 1.5 times the speed of the light wave they carry and the means free path between particles is the same length as the diameter of the Bahr atom. These particles possess most of the matter and carry nearly all the energy in the Universe. This energy is in the form of inertial energy of their own motion.

Rule number (4) The inertial forces that matter is capable of exerting is the only true force in nature.

Rule number (5) The atom has been redesigned. It is very small. It is only the diameter of the so-called nucleus of the atom. It is made up of protons and neutrons, which consist of free space electrons at **rest.** The proton as well the neutron is held together by the inertial presser of free space electrons. Clusters of various numbers of protons and neutrons are also forced together by the inertial pressure of the free space electrons. This is how the various elements of matter are formed. Outside of this cluster of protons and neutrons is a whole Universe of inertial energy carrying free space electrons which hold the cluster together and do all the things the different atoms do!

Rule number (6) Matter cannot be created or destroyed. When a neutron is smashed and broken up into its individual building blocks "electrons" they fly out into the surrounding space. Here the powerful inertial force of the free space electrons try to disperse them evenly in space, causing a tremendous outward movement of **inertial** energy that is equal to $E=MC^2$. Matter was not destroyed. It was only a change of state, solid matter changing into free space electrons.

Rule number (7) Most or possibly all physical phenomena are caused by a pressure difference.

Rule number (8) All matter is composed of particles commonly called electrons.

Rule number (9) The only form of energy Is the motion of matter. Most of this motion is possessed by the free space electrons.

I feel it is finally time to end this book. Writing Matter and Energy has been a flustering joy for many years. Pondering the unanswered questions of natures and proposing good theories to solve the mysteries has been a big part of my life for more then fifty years. The force that pushed me was primarily to find out how things worked.

(ONE PARTICLE)+(ONE FORCE)=(ONE UNIVERSE)

0-595-31900-9